MORE SCIENCE FOR YOU 112 ILLUSTRATED EXPERIMENTS

BOB BROWN

TAB BOOKS
Blue Ridge Summit, PA

To my great-grandchildren; Robbie, Pamela and Dustin

FIRST EDITION
THIRD PRINTING

© 1988 by **TAB Books**.
TAB Books is a division of McGraw-Hill, Inc.

Library of Congress Cataloging-in-Publication Data

Brown, Bob, 1907 –
     More science for you : 112 illustrated experiments / by Bob Brown.
          p.   cm.
     Includes index.
     ISBN 0-8306-9125-1   ISBN 0-8306-3125-9 (pbk.)
     1. Science—Experiments—Juvenile literature.   2. Physics-
– Experiments—Juvenile literature.   I. Title.
Q164.B8423   1988
507′.8—dc19   88-22744
                                                      CIP

Line drawings by Frank W. Bolle.
Technical assistance by Arthur Wood.
Cover photographs by Bob Brown, Jr.
Associate editor: Barbara Brown

Children pictured on the front cover (clockwise from top right) are; Aaron Allison, Sissy Rose Brulotte, Benjamin Moooley, and Sissy Rose Brulotte.

Contents

Notice

For children's use only with adult supervision.
Use chemicals or electricity only under adult supervision.
Use fire only with adult supervision.
Keep chemicals off of skin. Wash chemical apparatus.
Follow directions. Be careful. A chemical laboratory at home can be dangerous, but it can be lots of fun.
Chemicals can be poisonous even when absorbed through the skin.
Any cut in the skin should be washed immediately with plenty of water.
Don't work alone. Have someone with you, ready to turn off a switch if the experiment goes wrong—and this can happen with the best of us.

Introduction

At what age should children be introduced to science?

We are exposed to science from the cradle to the grave. Our daily lives are controlled by great and simple scientific truths. Our daily bread is made by scientific formulas. Our food is grown by scientific methods.

Watch a 3½-year-old play with small magnets. Observe the look of concentration on his face as he stacks the magnets on the floor and watches as they repel or attract, or as they swing in the air, one to another. Magic? Yes, magic to the child and also to the sophisticated adult who might like to take a peek as the phenomena change from magic to scientific reality.

This book contains experiments with vision, motion, weight, surface tension, temperature, sound, light, and many other concepts. We hope young people enjoy them.

Chapter 1

Experiments with
Vision and Motion

PROBLEM:

Magnify with a bowl.

NEEDED:

A clear glass mixing bowl with a rounded part and some water.

DO THIS:

Put a little water into the bowl, and examine something by looking at it through the water in the rounded part of the bowl. Magnification may be seen.

COMMENT:

This will not be a conventional lens, but light rays coming through it will be bent somewhat as they are bent as they come through a magnifying glass. This water "lens" may be focused if moved closer to the object being examined, or farther away.

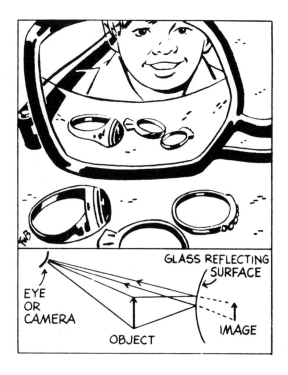

PROBLEM:

Sunglasses as convex reflectors.

NEEDED:

Sunglasses and small objects.

DO THIS:

Lay the glasses and objects on a table. In the reflection the objects will look smaller.

COMMENT:

The drawing shows the paths of some of the light rays that produce this effect. If the lenses of the glasses are flat, the glass acts as a simple mirror.

Five collections of these scientific tricks are available in book form. Get a free list from Bob Brown, 20 Vandalia Street, Asheville, NC 28806. Send a stamped self-addressed envelope.

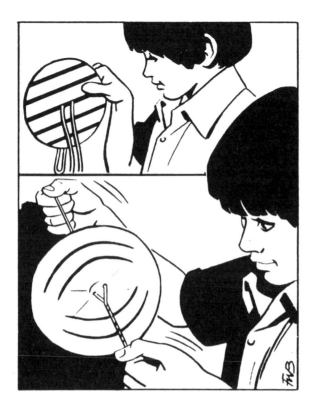

PROBLEM:

An optical illusion.

NEEDED:

A cardboard square or disk, crayons, string, a ruler, a compass.

DO THIS:

Mark heavy black lines across the cardboard. Make holes at the center and run the string through so the cardboard may be twirled by pulling and relaxing the string. As the card is twirled, the straight lines will appear curved.

HERE'S WHY:

This is an example of optical illusion, where the brain cannot interpret exactly what the eye may see. When rotating, the lines appear as circles.

PROBLEM:

Are the clouds moving?

NEEDED:

A mirror and a day when there are some clouds.

DO THIS:

Look up at a cloud and it may not appear to move at all. Place the mirror on the ground and look at the cloud through the mirror's reflection. Slight movement not formerly observed will be seen.

HERE'S WHY:

There is usually some wind in the upper atmosphere, causing the clouds to move. When large clouds move slowly and all together, they may appear not to move, and it is easy for the eye to move with them. When you look through the mirror, even a slight motion of the clouds will be observed because there is a background, the edge of the mirror, that is not moving at all.

PROBLEM:

The great tie-up.

NEEDED:

A long light rope, a small weight, plenty of space.

DO THIS:

Twirl the rope and weight around over your head until the rope is almost level. Then let the rope down so it begins to wind around you. It will make faster revolutions until the weight at the end snaps against your body.

HERE'S WHY:

An object in motion continues at the same speed unless the speed is altered by an external force. If the weight continues at the same speed as the rope becomes shorter, it must make faster revolutions. This is what takes place.

To show the winding rope better, the artist, Mr. Frank Bolle, drew three arms and three ropes. Only one person and one weight and rope are needed.

PROBLEM:

Make dry land ships.

NEEDED:

Paper boats, small magnets, a card table, a large magnet.

DO THIS:

Make the paper boats. Place them on the table away from metal. Put small magnets in them. As the large magnet is moved around under the table, the boats move about.

HERE'S WHY:

Magnets attract or repel each other, and the magnetic lines of force can penetrate the cardboard of the table top.

Try this with nails in the boats instead of magnets. It should work well.

PROBLEM:

Move the chair.

NEEDED:

A straight chair on the floor.

DO THIS:

Sit in the chair. Without touching the floor with the feet, make the chair move across the room.

COMMENT:

Friction between the floor and chair legs tends to hold the chair stationary as a jerking movement is started. When the jerking stops, the friction does not overcome the inertia, and the chair moves slightly. Continued jerking can move the chair across the floor, usually one side of the chair then the other.

PROBLEM:

The Doppler effect.

NEEDED:

An alarm clock tied on a long cord.

DO THIS:

Start the clock ringing, then whirl it around and around your head. The pitch of the sound will get lower as the clock goes away and higher as the clock swings again toward you.

HERE'S WHY:

Sound waves come from the clock faster as the clock moves toward the listener and are "stretched" as the clock moves away in the backward swing of the cord.

This is the Doppler effect (after Christian Johann Doppler, Austrian scientist), and is a common event. A car motor emits a higher pitch of sound as it approaches than as it goes away.

PROBLEM:

Running.

NEEDED:

Observation.

DO THIS:

Try running with the arms at the sides, then with arms moving back and forth. Why is it easier to run with arms moving?

COMMENT:

The upper part of the body usually moves as the legs move. If the arms are held at the sides there is little or no motion of the upper body and head. It is awkward and feels unnatural.

The action-reaction law comes into play with the motion of the arms; their movement helps propel the rest of the body.

Chapter 2

Experiments with
Weight and Surface Tension

PROBLEM:

Glamorize an old experiment.

NEEDED:

A plastic bottle or jar, cardboard, a container of water.

DO THIS:

Place the cardboard over the mouth of the jar, immerse the jar upside-down in the water, and if it is several inches below the surface the hand may be removed from the card. It will still cling to the jar mouth.

This is an old standard experiment. To make it different, squeeze a little air out of the jug. The card will now cling in less water depth.

WHAT HAPPENS:

Pressure upward on the card is greater than the downward pull of gravity, so the card stays against the jar. Squeeze a little air from the jar before placing the card on it, and the card will cling more tightly against the jar because the air pressure on top of the jar is lessened.

PROBLEM:

Surface tension (1).

NEEDED:

A pan of sudsy water, a jar, a cotton thread loop.

DO THIS:

Clean the jar mouth with soap or detergent, rinse it in clean water. Dip the mouth of the jar into the water, and draw it out carefully so a bubble film is left across it. Wet the thread, and place it carefully on the bubble. Touch the bubble inside the loop, and the loop springs into a perfect circle shape.

HERE'S WHY:

Surface tension is equal over the surface of the bubble film. Touching the film inside the loop destroys or weakens the tension there, and the tension outside the loop pulls the loop into circular shape.

PROBLEM:

Surface tension (2).

NEEDED:

A flat white plate, water, food color or ink, rubbing alcohol.

DO THIS:

Put a very thin layer of colored water on the plate. Get a drop of alcohol on the fingertip and touch it to the middle of the plate. The colored water will rush away from the finger.

HERE'S WHY:

Surface tension, like a thin sheet, covers the water. The surface tension of alcohol is less, so as the alcohol touches the water the greater surface tension of the water pulls the alcohol away from the finger.

PROBLEM:

Surface tension (3).

NEEDED:

A small plastic basket, a bowl of water, paper towel, waxed paper.

DO THIS:

Place the basket very carefully on the water, and it will float. Drop a square of waxed paper into the basket, and it still floats. Remove the paper carefully, and drop a square of paper towel into the basket. The basket will sink. (Berries and other produce come in the little plastic baskets at grocery stores.)

HERE'S WHY:

The surface tension film on the water is deformed by the basket but not broken. The water does not wet the waxed paper, and the surface tension is not broken. The water does wet the paper towel, breaking the surface tension at the basket edges. Because the basket is heavier than water, it will sink.

PROBLEM:

Surface tension (4).

NEEDED:

A drinking glass, water, cloth, a string or rubber band.

DO THIS:

Fill the glass with water. Wet the cloth. Stretch the cloth over the glass, turn the glass upside down, and note that the water does not run through the cloth and spill.

HERE'S WHY:

Surface tension seals the openings between the threads of the cloth, acting somewhat like a rubber sheet. This helps keep the water in the glass. Also, pressure of the atmosphere pushing up against the cloth helps keep the water in.

Try different kinds of cloth. Will open-mesh cheesecloth work? Does it take more than one thickness? Does it work better with the glass completely full or partially full?

PROBLEM:

Surface tension (5).

NEEDED:

A pan of water, metal jar lids, a hammer, and an 80-penny nail.

DO THIS:

Make two holes in each lid, driving the nail one way into one lid and the other way into the other. The lid in which the holes leave sharp projections upward will float. If the projections extend down into the water, the lid will sink.

HERE'S WHY:

Surface tension is like a thin plastic sheet on the water. If the surface tension is not broken, it should be strong enough to support the weight of the lid. The sharp projections break it, and the lid sinks.

PROBLEM:

Weight (1).

NEEDED:

Two scales.

DO THIS:

Stand with one foot on one scale and the other foot on the other. Will the sum of the two readings equal the reading if both feet stood on one scale?

ANSWER:

Yes. The weight of the person is the same in both cases, but is divided if two scales are used. Try standing on one foot on one scale.

(There are more than 1,000 tricky science experiments in Bob Brown's books. Get them from your bookstore, or get a free list from 20 Vandalia, Asheville, NC 28806. Send a self-addressed, stamped envelope.)

PROBLEM:

Weight (2).

NEEDED:

Iron nails, aluminum nails, two similar containers of water, a homemade balance with coat hanger and rubber bands.

DO THIS:

Hang the nails on the balance with the bands and string, and hold so the nails can be let down into water at the same time. The balance is lost.

COMMENT:

The iron will be heavier in the water than the aluminum because it displaces less water. Stated another way, the aluminum, while weighing the same in air, fills more space, or, has more volume and displaces more water.

Which is heavier, a pound of iron or a pound of feathers? If equal weights of iron and feathers are balanced at sea level then carried to a mountain top, they no longer will balance because the air high up is less dense than that at sea level.

PROBLEM:

Air pressure trick.

NEEDED:

Soda straw, a drink, a needle.

DO THIS:

Secretly, punch many small holes in the straw two inches from each end. Give the straw and a drink to a friend, and ask him how fast he can drink up. He should find it impossible to drink through the straw.

HERE'S WHY:

When we suck on the straw, we lessen the pressure of air on the part of the drink that is in the straw. The greater pressure on the surface of the liquid outside the straw forces the drink up.

When there are holes in the straw above the drink level, the pressure cannot be reduced because air comes in through these holes, and there is no pressure difference to force the drink into the mouth.

PROBLEM:

Action-reaction.

NEEDED:

Two people on roller skates; a cushion.

DO THIS:

Have the people toss the cushion from one to another, back and forth. Keep the skates parallel with each of the others, and it will be impossible for one person to stand still while tossing the cushion to the other.

COMMENT:

For every action there is an equal and opposite reaction. This is Newton's Third Law of Motion in action.

PROBLEM:

Water and air.

NEEDED:

Two wide-mouth fruit jars, round balloons, rubber bands, string, water.

DO THIS:

Fill one jar with water; leave the other filled with air. Cut the bottoms off two balloons, and slip the rest of the balloons over the mouths of the jars. Tie the rubber in place with rubber bands or string. Close the spouts of the balloons with string.

HERE'S WHY:

The balloon over air will stretch upward more than the one over water, because molecules in a liquid are close together with little space between them, and therefore resist compression or expansion. The air in the other jar can expand or contract, as is the nature of gases.

PROBLEM:

Pepper and surface tension experiments.

NEEDED:

Water, bowls, pepper, a spoon, vinegar, salad oil, milk, juice.

DO THIS:

Sprinkle pepper on the surface of water. The reason that it floats is because of a water skin we call surface tension. When you stir it in, it sinks.

See if surface tension holds up the pepper grains when the other liquids are used instead of water.

COMMENT:

Put pepper on the surface of the water. When you look carefully, you will see the grains sink slowly as they get soaked. When you put detergent in the water, the surface tension will break and the grains will sink faster.

PROBLEM:

The rising straw.

NEEDED:

Cold carbonated drink in bottle, a soda straw.

DO THIS:

Place the straw in the freshly-opened drink and it rises halfway out. Try this with a similar bottle of water, and the straw will stay down.

HERE'S WHY:

Carbon dioxide gas comes out of the solution and attaches itself to the straw in the form of bubbles. The bubbles make the straw lighter. Sunken ships are raised by this principle by attaching bags to the hull then filling them with air.

PROBLEM:

Weighty tricks.

NEEDED:

A bathroom scale.

DO THIS:

Stand on the scale. Raise your arms—you weigh more for an instant. Squat or bend over—your weight momentarily decreases. A greater change in weight occurs if two small pails of water are raised or lowered.

HERE'S WHY:

It is one of Newton's laws: every action has an equal but opposite reaction. When you raise your arms, you raise weight and your feet press down with an equal amount of weight on the scale platform. As you lower your arms, the opposite happens.

Chapter 3

Experiments with
Heat and Cold

PROBLEM:

Fire.

NEEDED:

A small piece of board, a cigarette, a kitchen match.

DO THIS:

Place the cigarette and the match side by side on the board so a third of them stick out beyond the edge of the board. Light them.

WHAT HAPPENS:

The match will be extinguished when its flame reaches the board. The cigarette will continue to burn until its tobacco has been consumed.

HERE'S WHY:

The match will be extinguished when the board conducts heat away from the flame so there is not enough heat left to promote combustion.

The cigarette does not have so much of its mass touching the board, and so cannot lose so much heat by conduction into the board. Also, a cigarette contains a chemical that promotes burning. This is convenient for the smoker, but makes the cigarette dangerous as a fire hazard. A burning cigarette has been called "an incendiary bomb."

PROBLEM:

Weather: rain.

NEEDED:

A saucepan, a stove, a dinner plate.

DO THIS:

Put half a cup of water into the pan, place it on medium heat, and cover with the plate. Lift the plate occasionally and note that drops of water are forming on its underside.

WHAT HAPPENS:

As the water is warmed, some of it becomes vapor. The vapor becomes liquid again when it touches the cool plate. The drops will get larger until they fall back into the pan.

This experiment shows, incompletely, how rain is formed. Water from the earth rises as vapor and condenses when it reaches cold air. It then can reform into liquid and fall as rain.

If a Pyrex pan is used, the action may be seen without lifting the plate.

PROBLEM:

A study in humidity.

NEEDED:

Two washcloths, water.

DO THIS:

Wet the cloths, wring them, and hang one in the room and the other in the refrigerator. See which one dries first.

COMMENT:

The speed of drying depends on relative humidity, which involves both temperature and amount of water vapor in the air. The low temperature in the refrigerator probably will result in higher relative humidity and slower drying. This should be tried other ways: hang one cloth inside and another outside on a cold day and on a hot day, a sunny day and cloudy day. Try it at night.

PROBLEM:

Freezing ice cubes.

NEEDED:

Ice cubes, pieces of thread.

DO THIS:

Make sure the cubes are just out of the refrigerator and very cold. Tie strings around them so they hang straight down. Let them come together very lightly, and they should stick together.

HERE'S WHY:

When the cubes are removed, the warm air of the room melts a little on the surfaces. This water refreezes when the cold cubes are brought together.

PROBLEM:

The foil.

NEEDED:

Kitchen foil, two thermometers.

DO THIS:

Wrap the thermometer bulbs with foil, one with the shiny side out and the other with the dull side out. Place them under bright light, and see whether one heats faster than the other.

WHAT HAPPENS:

The shiny side should reflect more light than the dull side, while the dull side absorbs more light and heat. Therefore, the thermometer wrapped with the dull side out should heat faster.

This can be applied to cooking. The potato wrapped with the dull side out should cook faster.

But does it actually work? There may be little or no difference in the thermometer readings. Could it be that the dull side of the foil next to a thermometer emits more heat and light than the shiny foil around the other thermometer? Experiment a little.

PROBLEM:

Why crush the ice?

NEEDED:

Ice cubes and crushed ice, containers of water, thermometers.

DO THIS:

Put about equal weights of crushed ice and cube ice in equal glasses of water. Note that the water with the crushed ice cools faster.

HERE'S WHY:

Ice, to melt, must absorb heat, in this case from the water. The crushed ice offers much greater surface area to touch the water, and so the heat transfer from water to ice is faster.

PROBLEM:

Hot or cold?

NEEDED:

A drinking glass, a neck scarf, and a refrigerator.

DO THIS:

Place the glass and scarf in the refrigerator. Leave them a while, then take them out. They both will be the same temperature, yet the glass will feel colder.

HERE'S WHY:

The sensation of cold comes as heat leaves the hand. Less heat can leave the hand holding the scarf because the woven scarf does not conduct the heat as the glass does.

A thermometer can be held around the scarf and glass with rubber bands. It will show that the temperature of the scarf and glass are the same as they are taken out of the refrigerator.

PROBLEM:

Mist from the pond.

NEEDED:

A tall, clear glass receptacle (a small fish bowl should do) and a cup of hot water.

DO THIS:

Place the cup or glass in the bottom of the receptacle. Watch fog rise from it.

COMMENT:

Fog is often seen rising from a pond or lake. The same explanation applies here. The hot water begins to evaporate, and as the vapor reaches the cooler air above it, the water condenses out into droplets. The air above the cup cannot hold as much water as the warmer air.

The fog produced in this experiment is not very thick. Look closely! (Try leaving the fish bowl in the refrigerator for a while before beginning the experiment.)

PROBLEM:

Cracking ice.

NEEDED:

Ice cubes, a soft drink.

DO THIS:

Drop ice cubes into the drink. A cracking sound may be heard, then a sizzling sound.

HERE'S WHY:

The drink will have a higher temperature than the ice. As the ice begins to melt, the difference in temperature between the surface of the ice and the interior produces stresses that may actually make small cracks. Very cold ice should be used.

The sizzling sound is caused by the bursting of bubbles when there are bubbles in the ice or when gas in the drink makes bubbles that burst.

PROBLEM:

Expansion.

NEEDED:

A tin can, a big nail, a hammer, pliers.

DO THIS:

Leave the nail in the freezer until it is very cold. While it is cold, drive it into the can quickly. Holding the nail with pliers, heat the nail over the electric or gas burner, then see if it can be pushed easily into the hole already made, without enlarging the hole.

WHAT HAPPENS:

Most metals expand when heated. In this case, the nail expands so much that it no longer fits the hole.

The most common metal that expands when cooled is type metal.

This is one of those experiments where the result is not spectacular. The expansion of the nail is very slight. I used an 80-penny nail.

PROBLEM:

Pop-bottle fog.

NEEDED:

A chilled, unopened bottle of pop.

DO THIS:

Notice that a thin fog forms at the mouth of the bottle when it is opened.

WHY?

Some of the gas in the drink expands when the pressure is reduced, and as it expands it cools. This makes some of the water vapor in the mouth of the bottle condense into fog droplets.

This is one of those experiments that everyone has seen, but that cannot be made to work every time. Many factors, such as temperature, humidity, and motion of the air, determine whether the fog forms.

PROBLEM:

Boiling over.

NEEDED:

A cooking pot on the stove.

DO THIS:

Note that the liquid boils quicker and may boil over if there is a lid on the pot. Remove the lid and the "boiling over" stops, if the pot isn't too full.

HERE'S WHY:

Considerable heat is lost to the air when the boiling takes place when there is no lid; therefore the boiling is not as vigorous above the surface of the liquid. The lid holds much heat in the pot, and the motion of air above the pot breaks apart many of the steam bubbles that cause the boiling over.

PROBLEM:

An electric fan.

NEEDED:

An electric fan, a thermometer, moist gauze.

DO THIS:

Place the thermometer in front of the fan and turn the fan on. Note that the temperature is not lowered. In fact, it may rise a little. Wrap wet gauze around the thermometer, hold it in front of the fan, and note that the temperature is lowered.

HERE'S WHY:

An electric fan does not lower temperature; heat from the motor can actually increase it. But the breeze can hasten evaporation from the gauze, and this will lower the temperature.

A fan cools a person not because it lowers temperature, but because it hastens evaporation of sweat from the body.

PROBLEM:

Heat conduction.

NEEDED:

Aluminum nails, iron nails the same size, a small board, flat ice.

DO THIS:

Drive some nails into the board, two iron nails at one end and two aluminum nails at the other. Place the nails and board on the ice.

WHAT HAPPENS:

The aluminum will conduct heat to the ice faster than the iron, and so the aluminum nails will sink into the ice more quickly.

It is interesting that in general, the metals that conduct heat best also conduct electricity best.

PROBLEM:

Evaporation.

NEEDED:

A jar, water, food color and ink, a jar lid.

DO THIS:

Put water and food color into a small lid. Put water and ink into another. Place them under a large jar, and put in a warm place (not direct sun) for a few days.

WHAT HAPPENS:

Water from both lids will evaporate, and the color, which is mostly solids, will remain in the small lids, perhaps in beautiful crystalline appearance.

Some solids *do* evaporate from the solid state. An example is moth balls.

PROBLEM:

A solar furnace.

NEEDED:

A concave mirror, black paper, sunlight. (Many make-up mirrors are concave.)

DO THIS:

Hold the mirror facing the sun, and adjust it so there is a very small bright spot on the paper. The spot can get very hot and begin to smoke.

CAUTION:

Do this outside, on the grass, and be ready to drop the paper and step on it if it catches fire. The mirror reflects the light coming to it from the sun and focuses that light into a small area on the paper. Don't hold the mirror so it might reflect into the eyes. Extremely bright light could damage the eyes.

PROBLEM:

The warm swim.

NEEDED:

Observation.

QUESTION:

Why does the swimming water feel warm after a rain?

HERE'S WHY:

Rain drops become cooler and cool the air somewhat by evaporation as they fall and actually are cooler than the body and the swimming water as they fall on them. Thus the swim water feels warmer by comparison. There is no illusion here, just a matter of relative temperatures.

PROBLEM:

Cold clothesline.

NEEDED:

Clothes on a line on a cold day.

WATCH THIS:

The water in the clothes can freeze, making the clothes stiff. Yet they will dry perfectly.

COMMENT:

One would think that the water in the clothes must melt before evaporating. But this is not so. Water can pass from the frozen solid state to the gaseous state without passing through the liquid state. This process is known as sublimation. A material that normally sublimes at room temperature is moth balls.

PROBLEM:

The steam iron.

DO THIS:

Try ironing a dampened cloth. Wrinkles will come out with steam that will not come out with a dry iron.

WHAT HAPPENS:

The hot steam moistens the cloth's fibers, making them limp and flexible, while the weight of the iron flattens the fibers, thus taking out some of their wrinkles. Some fiber wrinkles get so set in the cloth that even with a steam iron it is difficult to get them out.

PROBLEM:

What is boiling point?

NEEDED:

A candy thermometer, small pot, heat, water, salt.

DO THIS:

Boil some water in the pot, turn off the heat that makes the water boil, and put the thermometer in the water. Read the thermometer. Stir salt into the water, again bring it to a boil and see if the temperature is different.

WHAT HAPPENS:

Salt and most materials raise the boiling point of water. The salt water is thus hotter when it comes to a boil than plain water.

Altitude also affects boiling point. Water boils at a lower temperature atop a mountain than at sea level.

PROBLEM:

A flame.

NEEDED:

A lighted candle and a magnifying glass.

DO THIS:

Look carefully at the flame. You should see three parts. (Do not stand too close to the flame.)

HERE'S WHY:

Heat at the wick vaporizes some of the wax but does not support combustion, partially for a lack of air. The flame begins above this zone. At an even higher part of the flame, the flame's hot gases have mixed with air and are burning brightly.

WARNING:

Any fire is dangerous. Be careful.

PROBLEM:

The ice cube.

NEEDED:

An ice cube on water.

DO THIS:

Watch as the ice floats on the water. It will turn over.

HERE'S WHY:

This simple science trick can be made to look mysterious. The ice simply melts more on the bottom, where it is in contact with the warmer water. When it gets top-heavy, it turns over.

PROBLEM:

A living thermometer.

NEEDED:

Cricket (either wild or kept as a pet), a time piece.

DO THIS:

Assuming the temperature is between 45 and 80 degrees Fahrenheit, count the number of cricket chirps in 15 seconds, add 37, and there! The number is the temperature in Fahrenheit.

Many animals react in various ways to changes in temperature, barometer readings, and noises such as thunder. Moreover, people's moods also change with changes in weather.

PROBLEM:

Cool off.

NEEDED:

Observation in the shower.

DO THIS:

Take a lukewarm rather than a cold shower to cool off.

HERE'S WHY:

The sudden change from warm to cold contracts muscles and is not relaxing. A lukewarm shower removes excess heat and leaves the surface muscles relaxed.

PROBLEM:

The cool refrigerator.

NEEDED:

The kitchen refrigerator.

DO THIS:

Cool yourself by standing in the open door of the refrigerator.

WHAT'S WRONG HERE?

The refrigerator's compressor takes heat from coils in the food compartment and puts it into the coils closer to the air outside the refrigerator, where convection currents in the air take the heat into the kitchen. The heat coming from the refrigerator negates the cooling effect. Actually, trying to use the refrigerator as an air conditioner increases the temperature in the room. Heat from the refrigerator's motor is also added to the room air.

PROBLEM:

Dry ice.

NEEDED:

Ice from the freezer, a cloth.

DO THIS:

Wipe the cloth over the ice quickly when it is removed from the freezer. The cloth does not get wet. The ice in this case is "dry."

COMMENT:

Ice from a deep freezer has a temperature of about zero degrees Fahrenheit. It must warm up to 32 degrees before it starts to melt and get wet. This "dry ice" is not to be confused with frozen carbon dioxide, which is the real "dry ice." Its temperature is 110 degrees below zero Fahrenheit.

Chapter 4

Experiments with
Sound and Light

PROBLEM:

Television scanning.

NEEDED:

A pencil and a television set adjusted to bright.

DO THIS:

Hold the pencil upright and move it back and forth in front of the eyes as the eyes are focused on the TV screen. The pencil shows shadows at many places across the screen, and the shadows seem to show the pencil slanted although it is held upright.

HERE'S WHY:

TV pictures are formed by rapid motion of many horizontal lines moving top to bottom. As the pencil moves across the screen the shadow moves with the pencil and top to bottom at the same time, thus producing the effect of slanting.

PROBLEM:

Fireworks.

NEEDED:

A candle, some rosin.

DO THIS:

Put some powdered rosin in a pepper shaker and shake it over the candle flame. Bright sparks produced resemble "fireworks."

HERE'S WHY:

Rosin in a chunk does not burn readily, but when powdered the tiny bits burn rapidly because there is plenty of oxygen in the air around them.

Powdered rosin may be bought in a sports store, or chunks may be pounded into powder in a rag.

Try this with flour instead of rosin.

PROBLEM:

A stroboscope effect.

NEEDED:

A television set and a pendulum.

DO THIS:

Make the pendulum by attaching a small ball of kitchen foil to a string. Turn the brightness of the TV high, and the contrast low. Whirl the pendulum in front of the screen, and not one but several strings will be seen.

HERE'S WHY:

The light from the TV screen goes dim then bright several times a second. When it is bright, the eye sees the string; then the string moves to another place while the light is dim. Then the bright light comes on, and the string and ball are seen in the new location.

PROBLEM:

The plucked string.

NEEDED:

A guitar or other string instrument.

DO THIS:

Pluck one string with the finger, the fingernail, a pick, and with a gloved finger. Note the fine difference in the sounds.

HERE'S WHY:

The note produced should be the same, but it is not composed of only single vibrations. The various substances leaving the string produce other vibrations, called overtones, and these give the sound its difference.

The same note can be played on different instruments, but the overtones give each instrument its distinctive sound.

PROBLEM:

Reflection.

NEEDED:

Two mirrors, sticky tape, a penny.

DO THIS:

Fasten the mirrors together by putting tape on their ends. Place them at right angles, with the penny between. Count the number of penny reflections. Change the angle of the mirrors, and more pennies can be seen.

HERE'S WHY:

The image of the penny is reflected many times from one mirror to the other. It may be more interesting if a small toy animal is used instead of a penny.

PROBLEM:

The sounding board.

NEEDED:

A comb.

DO THIS:

Rub a finger along the teeth of the comb. The sound will be weak. Touch the comb against a table, a door, a metal container, a wooden wall—then rub the teeth. The sound will be considerably louder.

HERE'S WHY:

The sound is soft when the teeth alone produce the sound vibrations. When the comb touches another object, the vibrating comb causes that object to vibrate. Combined vibrations of comb and object cause more air to vibrate, making a louder sound.

PROBLEM:

Bend light.

NEEDED:

A hand and a fluorescent light.

DO THIS:

Look at the light through fingers almost closed together. Fringes of light can be seen through the narrow slits between the fingers.

HERE'S WHY:

Light is thought of as always going in a straight line. Here we see that this is not true. Light bends around corners to a slight degree. The strange light lines are called "diffraction fringes."

PROBLEM:

Vibrations.

NEEDED:

A yardstick or meter stick.

DO THIS:

Hold the stick against a solid table and start it vibrating by "plucking" it. As more of the stick is allowed to extend over the table, the frequency of the vibrations is lowered. When the projecting part is short and the frequency is higher, the vibrations can be heard as tones.

COMMENT:

As the stick vibrates up and down, it hits against the table, making the tones. The shorter stick vibrates more rapidly and therefore has a frequency high enough to be heard as tones.

PROBLEM:

Light and dust.

NEEDED:

A darkened room on a sunshiny day.

DO THIS:

Adjust a window shade so only a narrow beam of sunlight comes in. (Do not look directly into the sunlight.) Dust particles will be seen in the sunbeam, moving about.

HERE'S WHY:

No matter how clean is the house, there is always some dust. The particles are mostly light enough to float in the air, and they reflect some of the light that falls on them. This is not pollution; such a small amount of dust is not harmful. In fact, it helps us to see in shady places because light reflects off the particles.

PROBLEM:

Hear heartbeats.

NEEDED:

A paper cup and a knife.

DO THIS:

Cut the bottom out of the cup. Hold the hole in the bottom to your ear and hold the other end to someone's chest and listen for the sound of his or her heartbeat.

WHAT HAPPENS:

The cup concentrates some of the sound waves, making them louder as they reach the ear. Actually no cup or other device is necessary to hear heartbeats if the ear is held against the chest and the room is quiet.

PROBLEM:

Cut the Air.

NEEDED:

An electric fan.

DO THIS:

Blow into the fan as it runs. You will hear vibrations. Turn the fan off, and as the fan blades slow down, the vibrations become louder, not weaker as might be expected.

HERE'S WHY:

The turning blades pick up the air stream as they turn. As they slow down, each blade takes a larger bite of air, causing a louder but lower-pitched sound.

Chapter 5

Experiments with Tricks and Magic

PROBLEM:

Witchcraft.

NEEDED:

Four people and a small table.

DO THIS:

Sit around the table. Place all hands on the table, with thumbs touching and little fingers touching little fingers of those sitting next to you. Have lights low, maintain quiet, tell the table to move a certain way, then concentrate. After a while, perhaps 15 minutes, the table will move.

COMMENT:

There are perhaps people who would believe the table tilting or moving is a manifestation of the supernatural. There was a time when many took this seriously, some dubbing it witchcraft.

Michael Faraday, in 1853, demonstrated that tables moved because people pushed them. But the pushing need not be voluntary; the concentration causes the muscles to move without guidance from the conscious mind. Perhaps there are still people who do not accept the scientific explanation. Let's hear from them! (Bob Brown, 20 Vandalia, Asheville, NC 28806)

PROBLEM:

Sugar as a cleaner.

NEEDED:

Greasy hands, soap, water, sugar.

THE STORY:

Mix a little sugar with the soap lather to better remove grease and grime from the hands.

COMMENT:

The sugar does not enter into any chemical action that would help to remove the grease. But the sharp crystals of the sugar might help to "cut" the grease. Grains of sand will do the same thing, but they are not soluble in water and so are more difficult to get rid of.

Five books of these experiments are available. For a free list send a stamped, self-addressed envelope to Bob Brown, 20 Vandalia, Asheville, NC 28806.

PROBLEM:

Tricky shadow.

NEEDED:

A lighted fluorescent tube, a stick, a light-colored wall.

DO THIS:

Hold the stick parallel to the tube, and it casts a shadow on the wall. Turn the stick so it is at right angles with the tube, and the shadow disappears.

HERE'S WHY:

The shadow disappears because light from the ends of the tube falls on the places where the shadow should be and cancels out the shadow.

PROBLEM:

An oil and water trick.

NEEDED:

Water, oil, newspaper, writing paper, and waxed paper.

DO THIS:

Place a drop of water on the three papers and see what happens to it. Try a drop of oil on each paper.

WHAT HAPPENS:

On the waxed paper the water will form a ball if it is not too large, pulled into this form because the surface tension tends to make the sphere the smallest surface area. Fibers in the newsprint will allow the water to be absorbed (in other words, the paper will be wet). The writing paper may absorb the water.

The oil will be absorbed in the writing paper and newsprint, and again probably not in the oiled paper.

PROBLEM:

Reaction time.

NEEDED:

A yardstick or meter stick, some friends.

DO THIS:

Hold the stick up, and have a friend hold a hand around the bottom of it, ready to grab it. Let it go. See how far it plunges through the friend's hand before he catches it.

COMMENT:

It takes a little time for the eyes to see that the stick is moving, a little time for the brain to get the signal, a little time for the signal to reach the hand, then a little time for the hand to grasp the stick. All this together is the reaction time.

PROBLEM:

Wiggle-worm.

NEEDED:

A paper-covered plastic straw and water.

DO THIS:

Push the paper down on the straw to the end. Take it off; it will be accordion-pleated. Lay it down and touch a drop of water to it. It will **wiggle** as it expands, in an amusing fashion.

HERE'S WHY:

The paper is porous. The pores absorb water, making the paper larger. The irregular expansion is seen as the paper wiggles.

(Seven books of these scientific tricks are available. Get a free list from Bob Brown, 20 Vandalia, Asheville, NC 28806. Send a stamped, self-addressed envelope.)

PROBLEM:

Crazy egg.

NEEDED:

A large glass of water, an egg, a stream of water.

DO THIS:

Place the egg in the water and direct a stream downward on it. The egg will rise to the top of the water.

WHY?

Jearl Walker in his book *The Flying Circus of Physics* says nothing more than a description has been written on this. The explanation may be the Bernoulli principle: a fast-moving fluid exerts less pressure than a slow-moving fluid. The water flows faster on the top of the egg than on the bottom.

Select an egg that is heavy enough to sink to the bottom of the glass. Try different types of water streams; some may not work.

PROBLEM:

Water magic.

NEEDED:

Baking powder, water.

DO THIS:

Notice that there is no chemical action in the baking powder until water is added to it. Then it fizzes, producing carbon dioxide gas.

WHAT HAPPENS:

The ingredients of baking powder are there all along, but they cannot get together for any action until water brings them into combination. The carbon dioxide does not come from the water; water is only the catalyst that starts the action. There are many catalyst-induced reactions in chemistry.

PROBLEM:

A color trick.

NEEDED:

Mercurochrome and a glass of water.

DO THIS:

Put a few drops of mercurochrome into the water, stir, then look at a bright light through the glass. It looks red. Now hold the glass the opposite way, to one side, behind the light, so the light is reflected from the glass. The color is green.

HERE'S WHY:

When the light passes through the solution in the glass, red passes more easily, with some of the green and blue being absorbed. When the light is reflected from the solution, some of the red is absorbed and more green is reflected.

PROBLEM:

The traveling salt.

NEEDED:

Salt, a drinking glass.

DO THIS:

Add salt to a glass of hot water until no more will dissolve. Set the glass aside for a few weeks. The salt will be above the water level and on the outside of the glass.

HERE'S WHY:

Evaporation of the water will begin to leave a little salt on the glass. Capillary action will carry more salt up and eventually the capillary action will have carried salt solution to the outside of the glass, where it evaporates, leaving the salt.

This is the principle of the salt garden where salt "grows" on stones or pieces of coal.

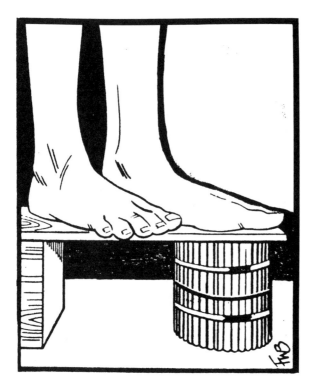

PROBLEM:

Strong paper.

NEEDED:

Corrugated paper, a small board, books or blocks, rubber bands.

DO THIS:

Cut the paper into even 5-inch strips and roll them together. Hold the roll with the bands. Place the roll on the floor, build up an equal height of blocks or books, and place the board across. The board will hold considerable weight. We could stand directly on the roll, but use of the board allows us to distribute weight on the roll evenly.

WHAT HAPPENS:

When paper is both rolled and corrugated, it is strong under compressionable force if the weight is evenly distributed.

PROBLEM:

Taste.

NEEDED:

Peppermint chewing gum.

DO THIS:

Have a friend close his eyes and hold his nose. Put the gum into his mouth and have him chew. He should not be able to tell the flavor of the gum. Have him stop holding his nose, then he can taste the peppermint.

HERE'S WHY:

Much of what we call taste is actually odor. This is true of the peppermint—it has odor but not taste. Simple, true tastes are sweet, sour, salty, and bitter.

PROBLEM:

Static and a bubble.

NEEDED:

An old long-play record, a small piece of fur or flannel, dishwashing detergent.

DO THIS:

Have someone assist by blowing bubbles from the detergent. Rub the record briskly with the fabric or fur, to give it a static charge, then hold it under the bubble as it is released. The bubble may be caught on the record, or it may float above the record, or it may be caught and rolled off the record.

HERE'S WHY:

The static electricity on the record may repel the buble, of course, the buble will have less or no charge, or perhaps it takes on a charge when it touches the record then it is repelled. Static is interesting!

PROBLEM:

Muscle-power electricity.

NEEDED:

A fluorescent lighting tube, plastic wrap, a dark room.

DO THIS:

Lightly rub the tube with the plastic. Part of the tube lights up as the wrap is rubbed back and forth.

HERE'S WHY:

The rubbing produces static electricity in the glass, enough to light the tube. Because the amount of energy is small, the light is dim. Do this in a very dark room.

PROBLEM:

A child's problem.

NEEDED:

A child in a swing on a rope, or a weight tied to the end of a string that is strung over a hook.

DO THIS:

Pump the swing or make the weight on the string "pump." How? Lift the weight by pulling down the other end of the string when the weight is swinging away from you and in the center of its swing. Stop pulling when the weight is at the end of its swing. This will cause it to swing higher. Never stand in front of the weight as it could hit you.

COMMENT:

A child does this experiment without thinking. Scientists can pore over much mathematics to explain it. Some of the physical quantities involved are energy, momentum, moment of inertia, and conservation of momentum.

PROBLEM:

Swallow.

NEEDED:

Food, water, a friend.

DO THIS:

Stand on your head and have the friend feed you food and water. Do this only if you are very experienced at standing on your head. See how easy—or difficult—it is to eat and drink without the aid of gravity.

COMMENT:

Gravity does help us to swallow our food, but the food is also helped down by a series of muscles in the mouth, throat, and esophagus.

PROBLEM:

The impossible nail.

NEEDED:

A nail, a hammer, a sponge or folded towel.

DO THIS:

Place the board on the soft material and try to hammer the nail into the board. You'll find it impossible to do.

HERE'S WHY:

When the hammer strikes the nail, the board moves with the nail because the soft material beneath it gives. The energy of the hammer is mostly used in moving the board. When placed on a hard surface, the board does not move and almost all the energy of the hammer is used in driving the nail.

PROBLEM:

A trick bottle.

NEEDED:

A bottle with tight-fitting lid, a transparent straw.

DO THIS:

Make a hole in the bottle cap and seal the straw in it with candle wax. Put some water in the straw and squeeze the bottle with both hands. Announce, "I can squeeze the bottle so hard it will make water rise in the straw."

HERE'S HOW:

The water can be forced up or out of the straw because heat from the hands causes air in the bottle to expand and to push the water out. If a rectangular bottle is used, however, it can be squeezed hard and the water will rise, because flat (not round) glass is elastic and will move a little when squeezed.

PROBLEM:

Little trick, big explanation.

NEEDED:

A thread spool, two weights, a string.

DO THIS:

Build the gadget as shown. Twirl the upper weight around and pull down on the lower weight. (Make sure that the upper weight hits *nothing* as it twirls.) The upper weight will twirl faster as the string is pulled down.

HERE'S WHY:

The speed at which the upper weight twirls depends in part on the size of the circle in which it twirls; the smaller the circles, the faster the twirling. Pulling down on the lower weight shrinks the size of the circle. Therefore, the upper weight will twirl faster.

PROBLEM:

Strips alive.

NEEDED:

Strips of newspaper about ½ inch wide and 8 inches long, a bowl of water.

DO THIS:

Fold the strips into accordion-style pleats. Press the creases tightly and drop the strips edge-first into the water. The strips seem to come alive as they move to straighten themselves out.

HERE'S WHY:

The strips absorb water quickly and their fibers begin to spread. The spreading tends to straighten the fibers out.

Chapter 6

Experiments with Miscellaneous Topics

PROBLEM:

Lung capacity.

NEEDED:

Two quart fruit jars, or one half-gallon-size fruit jar with capacities marked on the side, a short hose, an aquarium or large pan.

DO THIS:

Put water into the large container. Fill the jars with water and invert them in the large container so the water stays in. Hold the end of the hose under a jar, and take a deep breath of air. Blow all the air possible through the hose, and it will bubble up into the jar.

IT SHOWS:

How many pints of air the lungs normally hold. A quart jar may not be large enough, but two may be used, or one half-gallon size. Deep breathing regularly can increase the lung capacity. If a large container is available, a gallon milk jug can be used as shown in the drawing. The jug may be capped and the remaining water in it can be measured.

PROBLEM:

A yawn.

NEEDED:

Observation.

DO THIS:

In a quiet place where there are people, yawn when someone is looking at you. Chances are that person will yawn, too.

HERE'S WHY:

It is a matter of suggestion, and while suggestion may not be perfectly understood it is used all the time. Scratch behind your ear and see how many times someone will do the same.

Suggestion is an important factor in business, particularly in selling.

PROBLEM:

A raingauge holder.

NEEDED:

Two 1-by-2 strips 17½ inches long, two 36 inches long, four hinges, a screen door hook, wood screws, a gauge.

DO THIS:

Build the device as shown. The hook will hold it in the "up" position, and if a spring is handy, mount it as shown to absorb the jolt should the device fall.

COMMENT:

The author used this for mounting his rain gauge. A rain gauge must be above the roof level so blowing rain will not miss it. Without this device it would be necessary to climb a ladder each time the gauge is read. The gauge is about 37 inches lower when down than when up. Other uses will suggest themselves for this device. The gauge is available at stores. It is marked in inches and fractions of an inch.

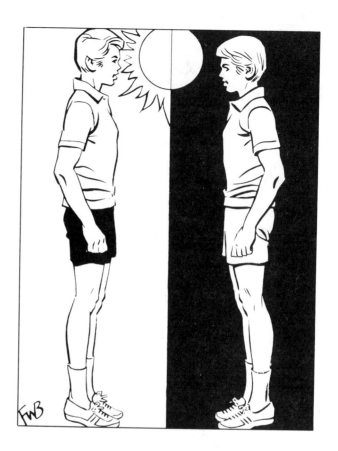

PROBLEM:

Taller, shorter.

NEEDED:

Exact measurement.

NOTE:

A person is a little taller when he first gets up in the morning.

HERE'S WHY:

When the body is stretched out, the fluids tend to be distributed in the body in a slightly different pattern. The spongy disks between the vertebrae soak up more fluid and expand, making the backbone slightly longer.

This added height is lost very quickly when the body again assumes the upright posture (suggested by Dr. I. Peterson, Kingston, Ontario.)

Send a self-addressed stamped envelope for a free list of the "Science for You" books. Mail to Bob Brown, 20 Vandalia, Asheville, NC 28806.

PROBLEM:

A sharpener.

NEEDED:

A clay flower pot, a knife to be sharpened.

DO THIS:

Hold the knife blade flat against the rim of the pot, move it back and forth, and it can be sharpened.

WHAT HAPPENS:

The baked clay of the pot rubs away tiny bits of the steel of the knife blade, because many of the particles in the clay are harder than the steel. If the correct particles are rubbed away, a keen blade edge is left.

PROBLEM:

Sharpen the scissors.

NEEDED:

Old scissors, kitchen foil, fine sandpaper.

DO THIS:

Remember the old superstition that scissors can be sharpened by cutting sandpaper? Someone has added "by cutting kitchen foil." Try both.

COMMENT:

Good scissors may be ruined by the sandpaper or foil. The sandpaper may add irregularities like saw teeth to the cutting edges. This may seem to make the scissors sharper, but it does not. Foil does nothing but perhaps make the scissors duller.

PROBLEM:

Static from tape.

NEEDED:

Sticky tape, cigarette ashes, other small particles.

DO THIS:

Pull the tape, hold it over the particles, and many of them will jump to the tape. Do not pull the tape entirely off the spool.

WHAT HAPPENS:

The friction created when the tape is pulled off the spool creates static electricity on the tape. In some kinds of tape the spool becomes negative and the tape positive, and in other kinds of tape the charges are the reverse.

Small pieces of styrofoam from a drinking cup are good for this experiment.

PROBLEM:

The drop as a measure.

NEEDED:

Waxed paper or a slightly greased pan, various objects, water.

DO THIS:

Dip a finger into the water, and let a drop fall from the finger to the paper. Let other drops fall from other objects: a toothpick, a spoon, a wash cloth, a medicine dropper, and others. Compare the sizes of the drops.

COMMENT:

The drop has been used for many years as a measure for medicines and other liquids such as flavorings. It is not a satisfactory measure, because drops falling from various objects vary too much in size.

PROBLEM:

Falling dominoes.

NEEDED:

Dominoes or bricks.

DO THIS:

Stack the objects as high as convenient. Lift the bottom one to topple the stack. They are not likely to fall together; there probably will be a break in the stack before they hit.

HERE'S WHY:

The top part of the stack would tend to turn less rapidly (angular acceleration) than the bottom part, but because both are attached this is not possible until stress develops to make the break in the stack. The sections then fall separately.

Dr. Jearl Walker in his book *The Flying Circus of Physics* says this phenomenon occurs when tall chimneys topple. They nearly always break as they fall.

PROBLEM:

Life of a tree.

NEEDED:

A stump or section of a recently cut tree.

DO THIS:

Observe the rings. There will be one ring for each year of the tree's growth. Rings on one side will be closer than on the other, closer on the north side because growth there was slower because of a lack of light and warmth. In a large tree, its leaves shade the south side, so the rings are wider east and west. If some rings are narrower than they should be, they represent slower growth in dry years.

PROBLEM:

Catching bugs at night.

NEEDED:

A two-cell flashlight and a dark, dry night.

DO THIS:

Hold the flashlight to your forehead and watch for bright spots in the grass or on the ground.

EXPLANATION:

The compound eyes of insects and spiders will reflect light as bright spots in the grass or on the ground. Bobby Nuckels, Scoutleader B of Iron Gate, Virginia, says that, with practice, you may be able to spot the bugs from a distance of 50 feet.

This must be done when there is no dew, because dewdrops will reflect the light.

PROBLEM:

A sponge garden.

NEEDED:

A dish, a sponge, some watercress, lettuce or radish seeds, water, a little water-soluble fertilizer such as Miracle-Gro.

DO THIS:

Using tweezers, drop the seeds into the openings in the sponge. Let the sponge rest in water on the dish.

WHAT HAPPENS:

Keep the experiment at room temperature. Water will rise in the sponge by capillary action and keep the seeds moist. They can make a small garden and will grow best if they have light.

PROBLEM:

Destroy cell membranes.

NEEDED:

Two potatoes, a knife, salt, a cooking pot.

DO THIS:

Boil one potato 15 to 20 minutes. Cut holes in both potatoes and place sugar or salt in the holes. Wait until the following day.

WHAT TO FIND:

Water in the hole in the raw potato, none in the hole in the cooked potato.

WHY:

Water and other liquids can travel through cell membranes, one to another, for considerable distances. This transfer is called osmosis. But for osmosis to take place the cells must be whole. Cooking destroys the cell membranes.

PROBLEM:

Bubbles.

NEEDED:

Bubble blowing solution, a blanket.

DO THIS:

Blow bubbles so they fall on the blanket. They may be made to roll around if the blanket is inclined.

HERE'S WHY:

Bubbles are expected to break when they touch any surface. But the surface of the blanket is made up of many points, and the surface tension of the bubble is sufficient to hold it together when it touches the many soft points of the blanket. When any considerable surface touches a bubble it is normally wetted at once and the bubble bursts. In this experiment the points are not wetted well when touched by the bubble.

PROBLEM:

Static cling.

NEEDED:

Styrofoam pieces and a woolen coat or sweater.

DO THIS:

Rub the styrofoam against the wool for several seconds, then throw it at an object. It should cling—unless the humidity in the room is high. In that case, no simple static experiments will work.

WHY THE CLING?

When the styrofoam pieces are rubbed against the wool they become charged, and hold the charge because they are poor conductors. When they come in contact with almost any object, they cling to the object because of the electrostatic force of attraction. They can cling because they are very light.

(The round styrofoam pieces shown are used as packing for merchandise shipping. Small pieces of styrofoam may be cut from drinking cups or ice containers.)

PROBLEM:

Why shiver?

NEEDED:

Observation.

COMMENT:

As the body loses heat, it cools. When arms and legs have lost so much heat that they are quite cold, the shivering starts. This motion of muscles is a form of exercise, and should have a warming effect. Is there a better explanation?

(Answered and unanswered questions are included in Bob Brown's books. Get a free list of them. Send a stamped self-addressed envelope to 20 Vandalia Street, Asheville, NC 28806).

PROBLEM:

Making butter.

NEEDED:

Whipping cream, bowl, mixer.

DO THIS:

Whip the cream as for whipping cream, but continue beating. It will form butter.

COMMENT:

This is called sweet cream butter. Butter also can be made from whole milk that has been allowed to sour. Fat in this case will separate to form butter, while buttermilk is left.

Making butter is not a chemical action, but a physical one. Salt and coloring may be added.

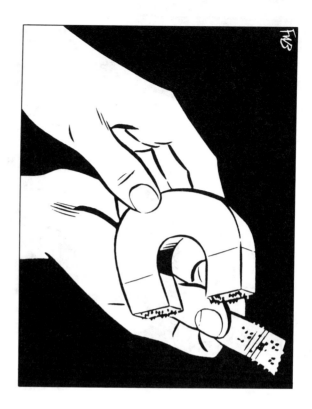

PROBLEM:

A helpful hint for the home lab.

NEEDED:

Sticky tape.

COMMENT:

Iron filings get on magnets and are difficult to remove. Sticky tape will get them off, and the tape with its filings can be thrown away.

Iron filings for home lab use can be obtained free from a friendly mechanic who turns brakes. They are useful in many experiments.

PROBLEM:

Popcorn.

NEEDED:

Popcorn, a measure, two plastic bags, oil, a popper.

DO THIS:

Measure equal amounts of corn into the bags. Freeze one bagful one day. Pop in equally hot oil and see if there is a difference in the end product.

COMMENT:

Someone on television stated that freezing the grains makes them pop larger. I tried this once only and found that the bowl of corn that had been frozen contained more corn. An explanation from readers would be welcome, but the experiment should be tried more than once.

PROBLEM:

Meringue.

NEEDED:

Egg whites and a beater.

DO THIS:

Beat the fluid until it becomes white and stiff. The change is commonplace but mysterious.

HERE'S WHY:

According to Jearl Walker, author of *The Flying Circus of Physics*, the protein molecules are in a spaghetti-like mass. Beating (or heating, as an egg is when fried) untangles the long molecules so they can attract each other to give a firmer structure. The color changes as the molecules rearrange themselves.

PROBLEM:

A better parachute.

NEEDED:

Very thin plastic such as from a dry cleaner's bag; scissors; string; a wadded piece of card for a weight.

DO THIS:

Tie the corners of a square piece of plastic to strings. Tie all four strings together with the weight. You have a parachute that can be thrown into the air without danger of damage from a heavier weight.

HERE'S WHY:

This was presented in my book *Science Treasures* some years ago. The use of plastic instead of a cotton handkerchief represents an improvement. It does not have to descend from a height to open.

PROBLEM:

Windshield cleaner chemistry.

NEEDED:

Rubbing alcohol, water, liquid detergent.

DO THIS:

Mix three parts alcohol, one part liquid detergent, and two parts water, and try using it as a windshield cleaner.

WHAT THE INGREDIENTS CONTRIBUTE:

The water dilutes the detergent, whose ions help to disperse particles of dirt by attaching themselves to the particles. This gives them a negative charge which makes them repel each other and dissolve in the water solution. The alcohol will help dissolve organic materials such as oil on the windshield.

PROBLEM:

Crystal beauty.

NEEDED:

Epsom salts, water, a mucilage type of glue, a pane of glass, a pan, heat.

DO THIS:

Heat the water and dissolve salts in it (a tablespoon of salts to a quarter cup of water.) Put in three or four drops of glue, stir, and spread the liquid over the pane of glass. (The glass should be clean.) A bit of paper towel may be used as a brush.

WHAT HAPPENS:

Place the glass where it will be undisturbed. As the water evaporates, beautiful needle-like, crystals form. Food coloring added to the water will make the crystals even more beautiful. The glue will make the solution spread over the glass.

PROBLEM:

A bone.

HNEEDED:

A pulley bone or leg bone from a chicken, a jar vinegar.

DO THIS:

Strip meat from the bone, place the bone in a jar of vinegar, and leave it three days. See if you can break it.

COMMENT:

The bone cannot be broken, it has become like rubber, because the acid vinegar has taken the minerals out. It is mainly minerals that give a bone its strength.

Soak a bone in a cola drink for a week and see if it loses its stiffness.

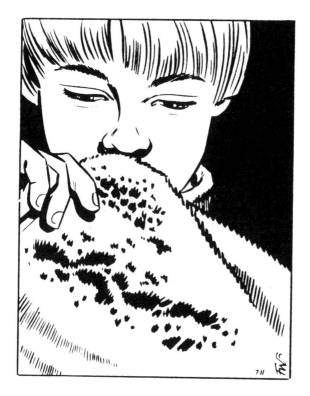

PROBLEM:

Mildew.

NEEDED:

Observation.

COMMENT:

Mildew is a fungus growth spread by spores which float through the air and begin growing on damp surfaces of organic proteins such as leather, which provide nourishment.

Mildew is difficult to control in many places. Keep surfaces dry and wipe frequently to remove spores before they get well established.

Mildew can often be removed from plant leaves by use of sulfur dust or commercial fungicides. The discoloration and odor of mildew in cloth or leather is often impossible to remove. Suggestions are invited from readers. Write Bob Brown, 20 Vandalia, Asheville, NC 28806.

PROBLEM:

The coiled hose.

NEEDED:

A garden hose on a reel, a funnel, water.

DO THIS:

Pour water into the end of the hose, as shown, and it is impossible to make the water come out the other end of the hose. Raise the end so the water pours into the hose from a higher point, above the length of the hose on the reel, and it may still be impossible.

COMMENT:

If the hose is full of water, and there is no air at the tops of the coils, water may be poured through.

Discussions of this are found in *Mathematical Games* by Martin Gardner and *The Flying Circus of Physics* by Jearl Walker.

PROBLEM:

Easy bubbles.

NEEDED:

A bowl and some dishwashing liquid.

DO THIS:

Fill the bowl with detergent and dip your closed fist into it. Draw your fist out, holding it so the tips of the thumb and index finger touch to make a ring. Bring the fist toward the mouth so air can be blown through the soapy film that is inside the ring.

Hold the palm upward; the fist will look like a funnel. Blow through the ring gently and large bubbles may be formed.

HERE'S WHY:

The liquid forms a soap film in the thumb-and-finger ring. When air is blown through the ring, the thin film is blown out of shape but holds together and closes to form a bubble. Surface tension holds the bubble together.

PROBLEM:

Onions and tears.

NEEDED:

Onions.

QUESTION:

Why does onion juice cause tears?

HERE'S WHY:

The allyl radical found in onion juice overstimulates and irritates the eyes. This produces the tears, which the eye excretes in an effort to wash away the irritant.

The allyl is found in oil of garlic and onions. The acid in onion juice is called propenylsul sulfenic acid. Don't forget this next time you peel onions!

PROBLEM:

Acid rain.

NEEDED:

Pink cascade petunia blossoms.

DO THIS:

Watch for white spots on the petals. Some say this bleaching is caused by acid rain.

Acid rain is known to affect some plants adversely. The acid is formed from gases in air from such sources as power plants burning coal and automobile engines burning gasoline. The extent of damage is being investigated.

PROBLEM:

A precipitate.

NEEDED:

Alum, ammonia.

DO THIS:

Dissolve a half tablespoonful of alum in half a glass of water. Add a little ammonia while stirring. A white precipitate, aluminum hydroxide, will form.

COMMENT:

This is a little kitchen chemistry. And the chemical name of alum? It is potassium aluminum sulfate.

Ammonia is ammonium hydroxide; therefore the precipitate is aluminum hydroxide.

PROBLEM:

Tree.

NEEDED:

A freshly cut tree, one gallon of water, a container, one bottle white Karo syrup (small), ½ cup of rubbing alcohol, a multi-vitamin-with-iron capsule.

DO THIS:

Make some cuts in the bark 3 inches up. Mix the ingredients and place the tree in them. It should stay fresh for a long time, according to an expert on "P.M. Magazine." The tree can absorb the liquid through fresh cuts in the bark, but not if the bark has dried out, closing the pores.

PROBLEM:

Rust.

NEEDED:

Three plastic dishes, nails or iron staples, vinegar, baking soda, water.

DO THIS:

Place staples in the dishes. Put plain water in one dish, put water and vinegar in another dish, and put soda and water in the last one.

WHAT HAPPENS:

In a few days, the staples in plain water will be still bright. Staples in vinegar and water will be rusted. Staples in the soda water may show some rust.

Vinegar will clear off the oily film that covers most staples, allowing them to react with oxygen in the air. Rusting is an oxidation process.

Index